# TAXING THE RAIN

By the same author:

*The Orchard Upstairs* (1980)
*The Child-Stealer* (1983)
*The Lion from Rio* (1986)
*Adventures with my Horse* (1988)

# PENELOPE SHUTTLE

# *Taxing the Rain*

Oxford   New York

OXFORD UNIVERSITY PRESS

1992

Oxford University Press, Walton Street, Oxford OX2 6DP

Oxford   New York   Toronto
Delhi   Bombay   Calcutta   Madras   Karachi
Petaling Jaya   Singapore   Hong Kong   Tokyo
Nairobi   Dar es Salaam   Cape Town
Melbourne   Auckland

and associated companies in
Berlin   Ibadan

Oxford is a trade mark of Oxford University Press

First published in Oxford Poets
as an Oxford University Press paperback 1992

British Library Cataloguing in Publication Data
Data available

Library of Congress Cataloging in Publication Data
Shuttle, Penelope, 1947–
Taxing the rain / Penelope Shuttle.
p. cm. — (Oxford poets)
I. Title.   II. Series.
PR6069.H8T38   1992   821'.914—dc20   92-4676
ISBN 0-19-282993-9

Typeset by Rowland Phototypesetting Ltd.
Printed in Hong Kong

*For Zoe and Peter*

# Acknowledgements

Some of these poems first appeared in the following magazines and anthologies: *Orbis*, *The Manhattan Review*, *Poetry London Newsletter*, *Verse*, *Empathy*, *Stand*, *Poetry Durham*, *Prospice*, *The Rialto*, *Skoob Occult Review*, *The Poetry Society Anthologies 1989/90* and *1991/92*, *The Orange Dove of Fiji*, and *Big World, Little World*.

'Breasts' was included in the travelling exhibition, *Making Waves: Six Poets for the Ninth Decade*. Poem Four of 'Sextet' appeared in a birthday symposium for Geoffrey Holloway.

The phrase 'No pudieron seguir soñando'—'They could not go on dreaming'—comes from Pablo Neruda, *The Unhappy One*, translated by Alastair Reid.

# Contents

## 'No pudieron seguir soñando'

'They could not go on dreaming.'

Storyteller North and Bystander South
stopped dreaming. So did East and West.

There was no more dreaming for Sun,
no more dreaming for Moon.

'No pudieron seguir soñando.'

No, they could not go on dreaming.
For all the Horses, no dreams.

For all the Mirrors, no dreams.
No dreams for the Forest,
despite his noble descent.

There is no tomorrow.
They could not go on dreaming it.

There's no tomorrow for the Mountain.
No tomorrow for the Ocean.

No dreams for the sober Whale.
No dreams for Mouse or Wolf.
No dreams for anyone.

Even the Street Women,
women of the poorer sort; they stopped dreaming.

No tomorrow for the Child
with his taste for solitude.

'No pudieron seguir soñando.'

Downhill Night and Tell-tale Dawn
stopped dreaming. So did Ghosts and Wheels.

No dreams for Globe-trotter Horizon,
no dreams for that drifter, Air.
No dreams for the Flower of Venus.

Lion stopped dreaming. Money
stopped dreaming.

All the Masks stopped dreaming.
Not one Razor or Scarecrow,
Web or Bone dreamed.

There is no tomorrow.
They could not go on dreaming it.
No.

The thinnest pane of Glass,
the most nimble and leaping Door,
the most devout Dog. No,
they could not.

Women stopped dreaming. So did Men.
'No pudieron seguir soñando.'
How could there be a tomorrow,

when none of the Trees dreamed?

When there was no dreaming for Snow,
for clannish Ant or veil-winged
Bee? Ask them to dream.

They cannot.

For even the dirty Water in the Gutter
has stopped dreaming. And not one of any
three Foxes can dream. No Thumb can dream
nor any of the Fingers of any Hand.

For there are no more tomorrows.

For the Ship cannot dream
and turns to cloud. For Cloud cannot dream
and rolls away like a stone
but it is no Stone.

The Invisible Man cannot dream.
The Seen Woman cannot dream.
Even for them, no tomorrow.

'No pudieron seguir soñando.'
They could not go on dreaming.

For the River, there's no tomorrow.
No tomorrow for the Dove,
faithful in marriage, chaste in widowhood.

Silver stops dreaming. Iron stops dreaming.
Pearls cannot dream. Nor can
Ice, Jade, Rice, Dew. No Heart
is dreaming. No Tongue is dreaming.

There is no tomorrow.
They could not go on dreaming it.

No one could.
Not even the Rain could.
Not even the Sky.

Whenever an earthquake occurs,
our planet rings like a bell;
but it has stopped dreaming.

Not one Stone in the World is dreaming.
The last word of all
is in your mouth;
slide it between my lips;
let me taste the last of salt.

## The Reader

Such secrets in books,
but they are all about you.
You read, you are the story,
you are the alien, the courtesan,
the murderer, the nun,
the son, the detective, the hero,
the victim.
You are in every primer, trilogy
and romance, you are on each greasy
page of the pornographies.
You read till you're shaking.
You read till your pockets are full of blood.
You read till your eyes are flying with tears,
you read till your thirst for knowledge
is slaked, and you try to come out
from between the pages.
Your eyelids close,
the book falls to the floor,
now you are just a big old sweet-tempered horse
called Annie,
asleep on your huge delicate pitch-dark hooves,
your great serene lips quivering;
the children reach up to stroke you
and the groom pulls coyly at your mane.

## Looking for Love

This woman looking for love
has to mend all the sails
that a winter of storms has ripped
to shreds, she will be sewing for years.

This guy must be looking for love.
See how he crouches like a runner
before the race.

Another man looking for love
crawls smiling up the phosphor mountains
of hell, the burning acrid slopes.

Two teenage girls speeding along
the towpath on golden-wheeled bikes,
are they looking for love?
Or are their shadows flying
over the naked Thames enough for them?

This woman looking for love
finds one tear in the rain.
If love were the remains
of a small wild horse,
or the skull of a young boy
lost in the wreck
of 'The Association',
she would have found love.

Two spitting cats
couple in the moonlit yard.
Is this love? thinks the peeping child.
She hopes not.

This woman looking for love
gets the door slammed in her face.
He's got too many wives already . . .

This fella looks for love
in the National Gallery,
he thinks he's Little Christ
snuggling and glittering on a Renaissance lap;
mother's boy,

whereas his wife,
rehearsing the Virtues and their Contraries,
finds love without looking,
in a room on the fourth floor of an ethereal hotel;
she doesn't ask his name.

## Taxing the Rain

When I wake the rain's falling
and I think, as always, it's for the best,

I remember how much I love rain,
the weakest and strongest of us all;

as I listen to its yesses and no's,
I think how many men and women

would, if they could,
against all sense and nature,

tax the rain for its privileges;

make it pay for soaking our earth
and splashing all over our leaves;

pay for muddying our grass
and amusing itself with our roots.

Let rain be taxed, they say,
for riding on our rivers
and drenching our sleeves;

for loitering in our lakes
and reservoirs. Make rain pay its way.

Make it pay for lying full length
in the long straight sedate green waters

of our city canals,
and for working its way through processes

of dreamy complexity
until this too-long untaxed rain comes indoors

and touches our lips,
bringing assuagement—for rain comes

to slake all our thirsts, spurting
brusque and thrilling in hot needles,

showering on to anyone naked;
or balming our skins in the shape of scented baths.

Yes, there are many who'd like to tax the rain;
even now they whisper, it can be done, it must be done.

## Home Birth

When the second small pale arm
is completed,
she casts the stitches off,
delicately lipping each one over its fellow,
neighbour over neighbour,
soft swallowing of stitch by stitch,
and the last stitch of all
has the pleasure and privilege
of having the wool passed through its heart
and pulled tight,
the end left to be threaded
into the seam
when she sews the separate pieces together.
Then the jacket will be folded up in tissue paper,
kept safe.
It waits for the birth of her child
one day next week in an ordinary household
among the sighs of her chaperones,
the shocked shuddering
of her stretched legs,
the suddenness of him, greasily
sliding out of her, his first wistful cry,
his sweet blunt-faced ugliness,
at home at once.

## Angel

The angel is coming down,
white-hot, feet-first,
abseiling down the sky.

Wingspan? At least the width
of two young men lying head to head,
James and Gary, their bare feet
modestly defiant, pointing north,
south.

The angel has ten thousand smiles,
he is coming down
smooth as a sucking of thumbs,

he is coming down on a dangle of breath,
in blazing bloodsilk robes.
See the size and dignity of his great toes!

He comes down in a steam of feathers,
a dander of plumes,
healthy as a spa,
air crackling round him.

Yes, he comes down
douce and sure-winged,
shouting sweetly through the smoke,
'J'arrive!'

Hovering on fiery foppish wings,
he gazes down at me
with crane-neck delicacy.

I turn my head from his furnace,
his drastic beauty . . .
he is overcoming me.

On my crouched back his breath's
a solid scorching fleece.
In my hidden eyes, the peep of him hurts.

He waits,
he will not wait long.

Flames flicker along my sleeve
of reverence as I thrust my hand
into the kiln of the seraph.

Howling, I shoulder my pain,
and tug out one feather,
tall as my daughter.

When I look up,
he's gone on headlong wings,
in a billow of smoulders,
sparks wheeling, molten heels

slouching the side wind. 'Adieu!'

Goodbye, I wave,
my arm spangled with blisters
that heal as I stare at the empty sky.

And the feather?
                    Is made of gold.
Vane and rachis, calamus
and down. For gold like an angel
joyeth in the fire.

## Afterlife

Dreamless since he'd gone, she depended on the day for love,
its intimidating brightness was better than nothing,
even if it wasn't really love,
even if that strange garden of his,
sunk in a long-necked big-bellied green glass jar,
translucent fernery thriving with myrtle, lichen and moss,
flourished more spectacularly since his death
than she cared to admit, as if within its eerie eden
some hope of his clung on, no, blossomed.

Then through the suddenly-transparent skin
of the pavement she saw her own husband faithful as ever,
lying as but one more of the dead laid out in their mute rows;
when the demons approached her with kind faces,
reasons and cushions, she pointed down there—Look!
Where he lay with all the others underground,
corked and cuffed in the terrible glassy light of that place.

And waking from her first dream after his death
she stumbles out of the bedroom of grief,
down into the early street
where the paving stones are solid, without revelation;
not one a window; she kneels, names him, but no voice
rises from the stone, no light scorches her face
or reveals the nature of the dead.

She lumbers up, stands relaxed, feet apart.
A neighbour's little dog barks tenderly,
he is what he is, the sky bears
its charmed blue life,
a flower-truck going from the farm to the city
dashes her with petals and hot exhaust, widow's confetti,
smell of jasmine and rose and smoke,
shocking her into laughter—
My Bob's aftershave! It smelt just like that!

# Breasts

Sleepily watching the ten o'clock news,
hearing the dull proud-voiced 'caster,
murders and madmen and bombs,
same old stories, routine famines in Africa;
suddenly through my drowse,
both my breasts burst out of my blouse,
lunging forward, buttons flying,
two individual active powers!
Yes, each vast breast had a life of its own,
they were like two huge tropical flowers,
and I was just their little overshadowed stem,
bearing two yearning creatures, enormous
giantess breasts!
I was surprised! But how I admired them,
I have to say it, they were courageous breasts,
each with a long sinewy walnutty nipple;
both breasts landed—thwack!
against the dusty tv screen,
longing to feed all those hungering children.
How I respected those breasts,
and when the cold glass cooled them,
banning their many-mouth-filling vocation,
I knew how bereaved they felt,
double-breasted hope gone,
and slowly, sadly those breasts of mine
deflated back to their normal size.
I touch them in despair.
Wasted children still gaze through burning air.
Remote and hunched, their dry mothers stare.
Drought wind toys with their brightly-beaded hair.

## Big Cat

A windowful of cloud.
Rain on the big sloping glass roof
falls from a once-only sky.

The lovers shiver,
their tongues spate
in their mouths
with a why and a how;
they say now, now.

The room holds them in its history,
between its pages,
as they tap at heaven.

They shake in their silver lining.

The window holds its breath.

Then the lovers come.
Then sleep purrs in their throats
like a big cat guessing names.

# Sextet

## One: My Ghost

Shoe of rain on right foot,
slipper of air on your left,
you're as neatly shod
as anyone I recall,
your hands warm
in their gloves of kisses,
your head buttoned
in its thousandth bonnet.
Whatever the weather
you are dressed for it.

Arthur and his Knights envy you,
yes, you.

## Two: Wish

The sky gift-wraps the world
in blue,
with cloud-ribbons and tags
marked Asia, America;
the world hangs
from a green branch
of some great decorated tree,
whose roots are set deep in living fire,
whose star-tipped crest
rises high and far and shining and forever;

a dream,
a fondness,
a thing I would make true;
here, with you. And with you.

## Three: The Nursery

Up,
the witch-broom
zooms up into the night sky
where it belongs,
the nobody astraddle its stem
nods in fiery reverie,
dark clouds brush her hair,
she sweeps the sky
on her long-handled broom,
the nobody
is off on her moon-hunt, flying
into the bewitching somewhere of the sky.

Her yelps of delight
are someone's lullaby,
someone sleeps soundly
in his mother's steadfast old crib.


## Four: Infant Prodigy

His morning honey is important,
his milk is necessary,
milk and honey, honey and milk.
The child has power and influence.
He is the speaker and the many-tongued singer.
His chair is high,
his gavel of orange plastic
is hung with bells;
they ring of luck.
He is the usual prodigy,
his wonder and ridicule
are spur of the moment; his fingers
over his eyes are science and philosophy;
his anger pardonable;
his love dignified and sanguine.
Breakfasted, he summons his woman.
She carries him wherever he wants to go.

## Five: John

*'Johnny-head-in-air'*

Nothing left
but his name

inscribed within
her ring,

his name ghosting
a child

learning to write
his own name;

always his father
is out-of-doors,

up high, like a moon,
writing across the sky

the same name as his,
John.


## Six: Spirits

*(School Play)*

Teacher breaks through a gap
in the scenery clouds;
glaring, she makes
the giggling white-robed infants
sing their song over and again,
their high voices squeaking like sweet chalks.

The tallest boy blows a trumpet,
the girls bang drums.
They must wake Scrooge.

'Remember,
all bring your tombstones tomorrow,'
shouts Miss, rallying, grabbing
a tearful child, buttoning
him swiftly into his oxfam anorak.

Then Jo's grown-up heavy-metal brother
carries him off
in a big leatherbound fraternal embrace,
away from the cardboard cemetery
and its daubed ghosts.

# Mademoiselle

She sleeps,
she is the star,
the geisha, the moth
called Knot Grass.
She is the long and short of it,
she is the blueness of the beetle's
belly, she is the ivory
fish, the clean-washed japanese
clothes. She is
the use of rainy weather,
the only valuable thing,
the surprising glove,
all the windows in Spain,
the free-faller.
She is also the man
riding in his dewy cart,
who comes to snow her
with januaries.
In her sleep
she goes wild-goosing,
at the far-fetched house
with the horse-shoe
hung on the door,
she calls out a name,
and who comes answering
from the spacious rooms
with their towering shadows?

# The Marriages

The newly-weds, the Marriages,
have new furniture in a new house,
naked rungs of chairs,
open eyes of mirrors,
no corners comforted yet by dust.
Lily and Paul are dancing shyly in a curtained room
to the incredibility of a flute,
zither and violin;
this is a night of starting and finishing,
this is a night of heart-shaped hours;
on sheets white as the astronomer's hopes,
their limbs are all thumbs
until suddenly—everything fits!
their voices are all devotion and slow cries;
they drift towards Orion
in an awe of sapphire, a gush of pearls,
the former Miss Lily and young Mr Marriage.

The dawn sky and the morning trees are bruised,
the bees are bruised,
the cats nudging across the yard
are bruised by love,
the rain is bruised by love; this
is as it should be, surely?

The Marriages have a rich and bruised look.
If you wish to live a happy life,
now is the day to begin, they agree, here in our house
where even the tables and chairs are happy
for the rooms are richly bruised by love
and the kissed carpets are hospitable and magic.

Mrs Lily Marriage makes the bed,
blushing, shaking pillows shyly,
quickstepping about the room,
while across town,
the young miller, the boyish baker,
in rolled-up shirt sleeves oversees
the manufacture of his loaves,
bending his head good-naturedly
before the rude casual jokes
of the white-capped floury-aproned women . . .
'How many loaves did you put
in 'er oven last evening, Master Paul?'
All shriek in saucy mock dismay . . .
though the more senior women pat his arm
curiously, as if touching for luck and courage
the sacred flesh of the newly-married Mr Marriage.

When he has drawn the trays of golden loaves
out of the deep ovens,
with their risen crusts and hot holy smell,
he bows to the ladies, playful and manly,
as befits the grinder
of Marriage's strongest finest flour.

## Neighbour

Holding the rain in your arms
you grow wiser, wetter!
The sky is not too big for you,
you could carry that too,
like a tree-bride or a hill-wife.
My house opens its eyes
and lets you give it tears;
how it weeps! and finds itself
next to the flooded garden.

You bring me this rain
that turns you to one
who knows the rain by heart,
whose sleeves turn to rain,
whose hands hold the rain
but as if by accident,
whose memory is now only rain,
whose future is rain,
who brings me the passions
and security of the rain.

You play the rain like a musical instrument.
Is it horn or string or percussion?
It is all. It plays, you play it,
music spills and drums, strums and hums.

As you play you observe me
with the eyes of rain,
my poorest neighbour,
my strangest friend.
You sing to me in rain,
you joke in rain,
you put rain into my arms
like flowers young as the hills.
How I hope the rain never stops,

for every inch of rain
surely is gifted
and deserves to be loved
as well as you love it.

# Rain's Child

On bad nights with baby
she uses the rain's voice
instead of her own.
But rain only knows what she knows.

The long thin damp green garden
runs down to the dawn river
like a lobbed carpet of welcome
in the wrong colour.

To soothe baby she uses the rain's voice
instead of her own.
But rain only knows what she knows;
that a child asks only

for simple nourishment and a name,
asks only that one
of a thousand possible doors
be opened;

that mother and rain
experience weariness and delight
in equal measure; the gift and pain
of the broken night.

The rain's child is not her child,
her child is not the rain's.

# Burn, Witch

Her loose puzzled jacket and boyish striped trousers recall her and our childhood as she sleep-patrols her house. Grace is sleep-walking. Or is Sleep grace-walking?

The kitchen smells of night, its manna; of moths and herbs; whiff of cat; nose-nipping reek of the well-scoured sink. Milky woody corners, and starched shirts hung on the door.

But she dreams her house is on fire. The white shirts blacken, tilt in meek smoky death off their twisting metal hangers.

Ouf! She steps easily over the red-for-danger toybox lying in wait. Even in the dark and sleep-walking, she's a mother.

Grace walks through her burning house, plumping up fiery cushions. She growl-smiles in a scalding mirror, sees her hair ablaze, crackling.

Where she walks, fire follows. Toys lift burning begging hands to her. Runtish inch-high warriors (fire-worshippers all) melt in black-hearted heaps. Robots blister, soldiers shrivel, giraffe and hippo and lion fry in their menagerie.

She stares at the smoky blur of family photos in redhot frames. She holds her feverish alchemic frog of bronze and crystal in her hand, blushing.

Sparks fly from the poor blind tv. She dances around her molten parlour. Burn, witch, burn! I will, she cries, yes, I will burn! And be not consumed.

Where once I brought hoovers, she cries, now I send flames! She waltzes around the ox-roast piano, the barbecue of books, the scorched brocades and plushes of chairs.

What I have created, I will destroy! Bless the cleansing flame! Grace leaps over the burning sofa, her flesh unharmed, though fire licks hungrily at the hems of new curtains.

The family photos sulk away to nothing. Grace goes barefoot up the fiery stairs, climbs back into bed, sliding between sheets of fire.

In comes her too-big-for-kisses son, sprawling over her side, muttering, he's dreaming too, he dreams he's a man, a saviour, he snatches the witch mother from her natural flames.

Grace lolls over his manly shoulder, while back in her dream, firemen are playing their big hoses on the flames, saving her hothead house again.

## The Peace

In medieval times,
in carvings of stone and wood,
church pillar, door arch and pew,
the richest joke
was thought to be

the sight of a husband
doing the housework.
In this misericord
from Ely Cathedral
a wife looks lazily and triumphantly

at her husband
pounding corn in a mortar;
woman's work.
In those days this image alone
was the very world turned upside-down.

With a relish
to equal the woman's,
two grinning mice
are hanging a cat
in this seat carving

from Malvern Priory;
satirically inverting
the natural order
and pre-empting
the peace of the Millennium.

# *Honeymoon*

Siesta. Mattress wide as a runway.
Our verandahed room white-walled, shuttered,
abruptly-furnished, not cool enough.
Iron-latticed grille in the door, peep-holing.

Outside, roofs blaze, tiles roast and crack.
We lie critical and naked in an odour of dust,
apricots, aloes. In the glass dish,
a litter of butts blurted out.

Sword point of sunlight
poking between the warp of shutters
scratches hieroglyphs on the far wall.
Squinting and dozing,
I can read its wavering script

but, jerked awake, forget.
(Outside, who's dinning murderously,
hammering nails into a heatwave hour,
too crazy to rest?)

Sheer heat hoods me in its nylon.
I blink and lick my lips;
my thighs are sticky,
on my throat, kissed bites flare,
half-slaked.

You snore; outside, a woman snarls, shrill and meek;
then silence, foreign, heroic.
You yawn in your sleep, as if eating fire;
sweat sheens and dulls on your arms.

This morning at the market
we bought bunches of young wet carrots,

soft black earth still clodded on them;
long green feathery leaf-fronds tickled my cheek;
a smell of neck napes, ear lobes, eyelids,
delicate flesh; lulled, I slide back under.

When I wake I stretch myself out full-length
beside your double, that man of sweetness,
who turns silently and browses above me,
aching into life, sliding in.

My hot lazy lips open, praising god and man,
while the sword of light
goes on spelling out what we can't read,
pricking and brailling the wall with its spiel,
its white-hot judder and stab asking
and asking and asking again.

## Good Dog

Why are the fish weeping?
Three-inch-long fish of bamboo green,
of black silk;
all the speckle-lipped fish weep.
Their tears are only a little crinkle
in the clear water of their tank,
like cellophane burning in sun.
Tears slide down their gawdy slinky barbels.

Why are the women weeping?
In the kitchen of sorrow
his wife weeps, her belly swells
big with fertile grief.
His daughter weeps, shrinking,
tiny tiny girl again.
His mother grasps her grief like a snake,
boldly, just behind the head,
shaking a venom of tears from its fangs.

Why is the world weeping?
Is it because the women weep so hard?
The roof leaks tears,
gutters grieve noisily,
the uncomforted sky floods the garden.

Next morning, the sun breaks through.
The women stumble up the bright mountain
of his funeral; husband, father and son.

In heartless heaven,
thinks his daughter, Daddy
is tracking angels with his favourite dog.

Angel feathers, thinks his wife, blushing,
are floating lazily around my love,
bristly and coarse as giant rhubarb leaves,
bright as the rainbow wonders of his fish.

My boy is taking heaven in his stride,
thinks his mother, he forgets what rain is,
what fish are, the names of women, our touch.

## Jesus

He drinks lakes, long rivers,
the little streams that belong to no one,
but he cannot quench his thirst.
He is the thirst
and he is the water of life,
and anyone who wants may drink him;
so his body must always
be disappearing underground
and welling up again, our lovely Jesus.
Women drink him,
they drink his waterfall heart,
he is their fountain and well,
he is the deeds and sufferings of water.
He is thirst, and yet they drink him!
Spendthrifts, not asking who this gift belongs to;
but swallowing the joy of it.
They lick up his flying colours of rain,
they lap from the open cup of his hands,
from the naked source of his side;
and the women get drunk on the Jesus-water,
each woman sighs and aches and is blessed on her bed,
she is floating on the beloved waters,
she desires to be wet, drenched, flooded,
as she comes, she gasps, 'Sweet Jesus!'

## Trick Horse

The trick horse is a bareback horse,
a holy horse
composed of many bare backs,
many couples in union,
near-naked men and women artfully entwined,
arranged in their copulations, loins
studded together so that their balancing bodies
create the body of a horse,

a yoga of sexy reciprocations.

The reason the trick horse brings good luck?
He's made from our inclination to fuck.

Men and women grip wrists, shoulder thighs,
weave, twine, dovetail, cling, clasp
and loll; are joined in the volupté
of their disguise. All support all.

It is an act of love, however they pose.
One woman's long swoop of dark hair—horsetail.

Yet the faces of these horse-makers
are as serious as any worshippers
in more austere religions; calm copulators,
the black-moustached men gazing about in sexual
reverie; the women, in gemmed bibs,
with pearl-studded noses, are just as grave

in their abandon. Though jewels of sweat
shiver on their golden skins, they assemble
as silently and devoutly as if in church
to make this estimable shameless mount.

One woman stretching her upper body forward
to make the horse's head,
(her glossed and outspread hair the mane),
grins and raises two hennaed index-fingers
to prick up the ears
of the animal of desire and delight.

Now comes the flower-garlanded rider
of this holy horse,
a girl thumb-belled, chime-toed,
a girl naked but for flowers and bells
and gold-braided bodice,
dancing towards her mount.

Now she lowers her baton of jasmine
and bows in greeting, the many eyes
of the horse watching her,
its many musks rising and clouding about her.

One man extends his linked hands,
making a stirrup for her gilded foot,
up she vaults,
with a leaping of bells,
scaling the carnal creature,
ascending her shivery gasping horse;
she squats and kisses the topmost balancing man,
who is persevering in his desire,
sinks herself down
upon her husband-saddle,
upon the pommel
of his stiff and risen phallus,
knees gripping,
flexing her haunches
till he's deep within her yoni;
and there she sits, radiant jockey,
astride, communing
with all the members of her love horse,
who quake in complicity and delight,
then steady, willingly taking her weight.

Now the girl rider, high upon her tantrick horse,
taps its composite flanks once
with her flowery rod;
this is the trick horse that all may ride,
that all may make.

She and her companions clip clop away in rapture.


In Indian erotic art the trick horse made of entwined balancing
lovers celebrates the power of sex and brings good fortune.

## My Moon

My moon goes everywhere.
My moon is happy because you are sleeping
and dreaming.
My moon is beyond pines and firs.
My moon is blue shadow under almond trees.
My moon has a flair for silence.
My moon plans the wedding day of honey and rain.
My moon loves to pilgrimage,
skimming hill and field
and coming home to me.

Dreaming of my moon,
I slide down her bosom of outstanding scenery,
her belly of whiteness.

My moon is sweet water I wash in.
I am always her last word.

# *Georgette*

No matter how often she moves the furniture
she can't find her Childhood.
It is named after the fashion of hurricanes,
Childhood Georgette.
But where is it?
No matter how often she coaxes old chairs
into new places, she can't find her Childhood.
Father grumbles quietly up and down the steps.
Mussed and sweaty,
she pushes everything back against the walls.
She looks and looks. Childhood?
Big hands clap the sky, Father is sending the rain.
She pushes her little foster bike through the rooms,
searching.
Father is at the window with his stormy thoughts.
He is shaking his branches.
Too old for a kiss? laughs Father out in the rain.
Again and again
he and the rain know what's right and what's wrong.
She leans the tear-stained bike against the wall.
Childhood?
A nudge of thunder. Don't tell lies!
She puts the furniture back how it was, everything
stares back obediently at her. Father
is muttering one of his old songs
and peeping round the door. Rain calls
out the name she never liked.
The windows don't lift a finger. Now Father
in those mirrors
is smiling at his little poupée.
Now she rides on Father's shoulders,
seeing everything, interpreting nothing.
Georgette is riding. Ice and rain on the stairs,
all the rooms galloping round and round
and hurting, Father knows where her Childhood is.

## Jesus is in love with Russia

Jesus leans down from his sky,
he elbows his moon aside,
he is in a rage with Russia,
he pulls Russia up from her icy knees,
hear her rivers creaking,
Neva and Volga, Don and Kama,
he is so angry,
like some secluded Czar, winter-palaced,
it is Russia that enrages him,
he is not on speaking terms with Russia,
angrily he licks the salt from her lips,
bitterly he stares in her ice-blind eyes,
he rubs his hands just as he pleases
over her brow and her freezing cheeks,
mouth to mouth he gives her
scorching wrathful kisses of life,
blasts his holy breath down her throat,
but she doesn't smile, or anything!
He shakes her by the shoulders
till all her snowy mists are dashing and flying.
How he yanks at her snows!
But Russia will not move for him,
will not wake or walk or care about him.

Unsleeping bad-tempered stubble-chinned Jesus
storms round her. What's wrong? he yells.
Do you want a more beautiful or mysterious name?
Is that it?

When will Russia take what he offers,
when will she use his anger and his power,
doesn't she know he expresses the divine mind?
When will she love him as he loves her?
Sulkily he curls up in Russia's lap,
soon he dreams Russia is singing to him
in all her golden churches.

## *Nuptial Arts*

Ssh. Don't talk. Listen.
Our bedroom smarts with summer lightning,
white cloth flicked off the sky table.

There! Thunder, saying its name,
my name, your name. Then the moon,
rolling in and out of her radiance.

Your cock's tame to my touch,
connubial, hot; my lips
ring it, my tongue hushing you.

Silently you trace out
the big white freak kiss of sun
X-ed on my tanned back,

then touch me lower, where the clouds
of me are all dew and shock.

Let us practise our nuptial arts.
Let us embrace and if when you come into me
our sighing cannot be heard

for the sweet growl of thunder overhead,
then let the sky speak for us,
conversing in both fire and rain.

## Isabella's Chair

A chair capable of Mozart—
He sat here! she insists.
Maybe.
Only the afternoon sits in Isabella's chair.
No one else cares to sit there,
despite Isabella's best satin-blue cushion,
fringed and tasselled, threaded with tadpoles
and roses.
No one likes the chair, its daily-dusted regality,
its disdainful veneered opulence,
its laze of upholstery, its swagger of brocade,
its fussy gilded feet snagging the old worn carpet.

In Isabella's house, something has to go
to make room for that chair.
She studies her rooms with the gentlest
of questions, the meanest eye.
Shall I surrender this table, or that couch?
Shall I banish the sofa? Or silly old Grandpa Clock?
She considers the future of her bed,
its lack of celebrity. And who needs a wardrobe
when such a chair honours my house?

Oh what a monarch of a chair; she loves him.
Soon she will dare to sit there like a consort.
Isabella puts on her new dress,
patterned with lilies and rain-bringing clouds.
In her finery she walks amid frail doomed furniture,
her cowards,
saying, this must go, and this must go.
For my chair must be in good company.

From the rough side of the curtain,
she watches the removal van drive away.
Now all her rooms are bare and free,
just a shine of dust lifting and drizzling,
just doors stooping over her,
mirrors offering friendship.
Her glance goes round and round
and rests lovingly on her chair.
She was aware of his secret sufferings,
surrounded so long by mortal furniture.
But that's all over now, she whispers.

In the big cool bare room, in catling twilight,
she tugs the chair carefully into place
before her favourite mirror.
She is bathed, she is talcumed,
she sees how perfect a chair awaits her,
his partnering mood is all devotion to her.

This is the chair I have chosen,
gasps Isabella, crouching in pain.
Installed in her chair of high favour,
she lifts her skirts, shivering.
The mirror sees everything
but only the chair understands her,
he is her intimate, her accompanist,
she digs her nails into his brocaded arms,
her sweat soaks into satin, velvet, gilt,
her blood runs down his seamed and stitched back,
he supports her, he bears her weight,
he is her chair of ceremony and privilege;
she cries out—
her son, webbed and filmed in flesh's strong fabric,
is born—in Mozart's chair!
Tiny Wolfgang sneezes and wails—music!

## Changing

Certain places in the garden are female,
others are male. It is like
the male and female milks.
The night comes here as a boy
of great boldness and purity.
Hidden among tongue-shaped leaves
that lick up the rain,
he changes to a woman.
She has the beauty of her nighthood,
she is a changer, she has the look
of magical necessity.
For her the moon lets down a little hot pisse.
For her the garden makes a once-only offer.
For her the sky takes to its bed.
For her the past comes running back.
For her the fig-tree laden with its yonis
slowly and unashamedly turns to a man.
Changing and changing,
he and she try all the colours, all the smells
of this certain garden.
They will not talk to you. You can only dream of them.

# Gift Horse

My gift horse has steep black wings
of birthday glass,
he has no saddle.

He has a mirror for his stable,
rears up on his blue-tile courtyard,
looks fondly round,

his eye clear and clever
as a little icon.

He can go anywhere by foot,
or by air, with his good manners.

He crops gardens
that tend themselves;
the seemly wildernesses.

He's gift and giver,
his wings salaaming over me;
he brings me the last leaf
from a lost forest.

He gives me a whispered Koran
and a globe for shaking
till a starland of snow falls,
too cold ever to melt.

He brings me slippers of rosy fur,
so that my swaddled feet blush.

He brings me a bowl of tiny round mirrors
showing me everything.

Sometimes my gift horse shudders
as if he loves me or the day too much;
so strong is the joy of the donor.

At the very moment of rain
he brings me news, who kisses whom,
who breaks her promise . . .

Then he prospers into the air
on his strong slim wings;
he is not the wandering Hyracotherium,
he is not any old galloping Mesohippus;
he is my gift horse.

He gives me all I need,
my Equus of black glass.

## She Sends

Moon pretends
to be surprised at the rain.
As if rain wasn't her middle name.
Rain knows all the words
moon hides
in her blue purdahs
and fans of cloud.
At the top of my voice
I find silence,
I mean,
*moon*.
Around me, the laden house
lists in its own quiet.
The uncurtained window
is a dark where moon
baffles my one lamp.
My needle winks like diamond
as I sew fast,
as if in heaven or hell.
Moon sighs crossly at such a task,
dares me to return
her gaze.
When I do not,
she sends a shadow man
to groom me,
mouth to mouth,
his technique shushed
as any underwater music.

# Boy Thief

## Day

At dawn he resumes his life of crime,
he nicks your best pencil, pickpockets

your lipstick, your keys,
he doesn't know any better.

He kidnaps your slice of breakfast melon,
spirits away book, ear-ring, glove.

He raids the laundry basket, odd socks
haunt the house for weeks.

He broods over his dirty work.
In broad daylight he lies in ambush,

then paws through bedroom cupboards,
snake-charming your stockings,

sniffing your perfume
with a shocked grin. He hides his loot

everywhere, a blue scarf from Japan
floats in the rainwater butt,

for days your new shoes wait,
reticent behind the washing machine,

your sun-glasses go into exile
in the cat's basket, a spoon-drift

glints behind the curtain,
the kitchen broom's moored in the rosebed,

watch out, there's a thief about.

## Night

Your boy thief robs you of sleep.
The whisper in the dark

that might tell you everything
is lost in his cry.

Your dreams?
He demands the lot,

midnight extortioner,
tantrum howler:

give me everything!
That laughing chair,

that room of clouds,
that feast! For he knows

power pours into the boy
who eats off a superman plate.

At last he sleeps,
sprawled in your bed,

cuddling your old fur cape,
his light fingers

leave you nothing,
not even the peace of your own pillow.

## Zoo Morning

Elephants prepare to look solemn and move slowly
though all night they drank and danced, partied
and gambled, didn't act their age.

Night-scholar monkeys take off their glasses,
pack away their tomes and theses,
sighing as they get ready for yet another long day
of gibbering and gesticulating, shocking
and scandalizing the punters.

Bears stop shouting their political slogans
and adopt their cute-but-not-really teddies' stance
in the concrete bear-pit.

Big cats hide their flower-presses, embroidery-frames
and watercolours;
grumbling, they try a few practise roars.
Their job is to rend the air, to devour carcasses,
to sleep-lounge at their vicious carnivorous ease.

What a life.
But none of them would give up show-business.

The snakes who are always changing,
skin after skin,
open their aged eyes and hinged jaws in welcome.

Between paddock and enclosure
we drag our unfurred young.
Our speech is over-complex, deceitful.
Our day out is not all it should be.
The kids howl, baffled.

All the animals are very good at being animals.
As usual, we are not up to being us.
Our human smells prison us.

In the insect house
the red-kneed spider dances on her eight light fantastics;
on her shelf of silence she waltzes and twirls;
joy in her hairy joints, her ruby-red eyes.

## Delicious Babies

Because of spring there are babies everywhere,
sweet or sulky, irascible or full of the milk of human kindness.
Yum, yum! Delicious babies!
Babies with the soft skins of babies, cheeks
of such tit-bit pinkness, tickle-able babies, tasty babies,
mouth-watering babies.

The pads of their hands! The rounds
of their knees! Their good smells of bathtime
and new clothes and gobbled rusks!
Even their discarded nappies are worthy of them, reveal their powers.
Legions and hosts of babies! Babies bold as lions, sighing babies,
tricksy babies, omniscient babies, babies using a plain language

of reasonable demands and courteous acceptance.
Others have the habit of loud contradiction,
can empty a railway carriage, (though their displeasing howls
cheer up childless women).
Look at this baby, sitting bolt upright in his buggy!
Consider his lofty unsmiling acknowledgement of our adulation,

look at the elfin golfer's hat flattering his fluffy hair!
Look next at this very smallest of babies
tightly wrapped in a foppery of blankets.
In his high promenading pram he sleeps sumptuously,
only a nose, his father's, a white bonnet and a wink
of eyelid showing.

All babies are manic-serene, all babies are mine,
all babies are edible, the boys taste best.
I feed on them, nectareous are my babies,
manna, confiture, my sweet groceries.

I smack my lips,
deep in my belly the egg ripens,
makes the windows shake,
another ovum-quake
moves earth, sky and me . . .

Bring me more babies! Let me have them for breakfast,
lunch and tea! Let me feast, let my honey-banquet of babies
go on forever, fresh deliveries night and day!

## Faire Toad

'Foul toade hath a faire stone
in his head',
especially is this fine
in the heads of old and great toads;
the fairer the stone,
the stronger his venom.

Until today I never knew
toads shed their skins; in muck and mud,
agony and triumph of transformation,
the magic of the amphibian.

Faire toad feeds only on living prey.
He carries his heart in his throat.
But for the mating time
he lives the life of a recluse.

He climbs higher and spawns deeper
than cousin frog.
When man and wife toad embrace,
they do not cease for hours
their amorous encounter.

A toad may choose to live in water or earth.
Great is his capacity for fasting.

No, he does not spit fire.
No, he does not love women,
though in stories he is royal
as any frog.
He does not suck milk from cows.

Scholar-toad sees more than we see,
his eyes are eight times
more sensitive to light
than our human clarity.
Such glooms his eyes can pierce . . .

My lady toad gives herself
in wax, iron, silk and wood
on the wayside altars of Central Europe.

Luck shuffles
in to the parlour which a toad visits.

If you wish for his treasure,
do not torment him or dress him
in new silk;
but give him warmth, kindness,
pleasure; then he'll be buffoon
and priest, hopping
and processing in his khaki, bronze
and sepia swarth . . . leave
you his jewel-head in his will,
a faire and blood-hot ruby.

# OXFORD POETS

Fleur Adcock
Edward Kamau Brathwaite
Joseph Brodsky
Basil Bunting
Daniela Crăsnaru
W. H. Davies
Michael Donaghy
Keith Douglas
D. J. Enright
Roy Fisher
David Gascoyne
Ivor Gurney
David Harsent
Gwen Harwood
Anthony Hecht
Zbigniew Herbert
Thomas Kinsella
Brad Leithauser
Derek Mahon

Jamie McKendrick
Sean O'Brien
Peter Porter
Craig Raine
Henry Reed
Christopher Reid
Stephen Romer
Carole Satyamurti
Peter Scupham
Jo Shapcott
Penelope Shuttle
Anne Stevenson
George Szirtes
Grete Tartler
Edward Thomas
Charles Tomlinson
Chris Wallace-Crabbe
Hugo Williams